MW01319057

MYSTERY EXPLORERS™

SEARCHING FOR CLOSE ENCOUNTERS WITH ALIENS

Ian M. Goldfarb
and Janna Silverstein

New York

Published in 2012 by The Rosen Publishing Group, Inc.
29 East 21st Street, New York, NY 10010

First Edition

Library of Congress Cataloging-in-Publication Data

Goldfarb, Ian M.
Searching for close encounters with aliens/Ian M. Goldfarb, Janna Silverstein.
p. cm.
Includes bibliographical references and index.
ISBN 978-1-4488-4761-7 (library binding)—
ISBN 978-1-4488-4769-3 (pbk.)—
ISBN 978-1-4488-4777-8 (6-pack)
1. Extraterrestrial beings. I. Silverstein, Janna. II. Title.
QB54.G55 2012
001.942—dc22

2011006533

Manufactured in the United States of America

CPSIA Compliance Information: Batch #S11YA: For further information, contact Rosen Publishing, New York, New York, at 1-800-237-9932.

CONTENTS

INTRODUCTION

Humanity has always wondered about life on other planets. Books and movies about contact with beings from outer space are compelling because they explore the idea that we are not alone in the universe. However, even fictional accounts are not as strange as some of the encounters that people report today. Alleged abductees describe lights in the sky, ships that travel at impossible speeds, and gray beings with big, black eyes. Dim recollections of frightening abductions sometimes haunt people for the rest of their lives. What are these people experiencing, and what is the truth behind their stories? And is there some kind of cover-up going on?

Skeptics contend that even if there is some kind of life elsewhere in the universe, the likelihood that it has evolved to an intelligent level is very slim. Even if we believed that highly intelligent life were possible, skeptics say, such a civilization would probably not be advanced enough to travel at the speed of light. And, if there were actually beings sophisticated enough to master star flight, they would probably not bother visiting Earth, a primitive, backwater planet on the edge of the Milky Way.

If all of this is so, then how do we explain the experiences of people like Betty and Barney Hill, Kathie Davis, Travis Walton, and the thousands of others who tell about their experiences with alien beings? Some stories may be hoaxes or dreams. But we can't explain them all.

One thing seems certain: something is happening to thousands of people around the world. Whether they are encountering demons, angels, leprechauns, simple nightmares, or alien visitors, they are experiencing something unique, something that will continue to capture our imagination and attention well into the future.

The hit film *Close Encounters of the Third Kind* influenced popular images of alien visitations.

CHAPTER 1

Era of Aliens

It all began on June 24, 1947, a clear, calm day. Kenneth Arnold, a deputy sheriff and businessman, as well as a trained private pilot, was flying his airplane around Mt. Rainier, about 90 miles (145 kilometers) south of Seattle, Washington. As he flew north, he spotted nine odd objects speeding in a V formation at what he estimated was over a thousand miles per hour. When Arnold described the objects he'd seen as saucer-shaped, the press jumped on the description and dubbed the crafts "flying saucers." A burst of flying saucer sightings in the United States followed this event.

In the early 1950s, the U.S. government chose to use the term "unidentified flying object" rather than flying saucer. An unidentified flying object, or UFO, was exactly that: an object flying through the air that was unlike any recognizable aircraft. Ever since this wave of sightings began more than sixty-five years ago, Americans have been intrigued by the UFO mystery and the possibility of aliens visiting from other planets.

The Roswell Incident

Only two weeks after Kenneth Arnold's sighting, rancher Mac Brazel discovered metallic debris spread over acres of his ranch just outside Roswell, New Mexico. Unhappy about the garbage, which he assumed to be military wreckage, he called officials at Roswell Army Air Field (RAAF), the local air force base. The RAAF issued a statement claiming to have recovered a crashed "flying disk." The local newspaper reported the story, and the news quickly spread around the world. The military initiated a high-security search and cleanup operation. Witnesses who examined the debris reported seeing pieces of metal with symbols thought to be writing and bits of foil that couldn't be burned, creased, or torn.

Soon afterward, the initial press release was retracted on orders from General Roger M. Ramey, commander of the Eighth Army Air Force, now in charge of the investigation. Major Jesse A. Marcel, who had been handling the research and recovery mission up to that point, stated that the debris was nothing more than the wreckage of a weather balloon. The quick change in the story fueled suspicions that the government was covering up the facts.

In the years following the crash, other strange evidence was gathered. Local witnesses testified that they saw debris they couldn't identify. Firefighters stated that they had seen a second crash site where shrunken, burned bodies

Something fell to Earth near Roswell, New Mexico, in 1947, inspiring decades of speculation about the crash of an alien spaceship. A UFO museum in Roswell explores the incident.

were found. There were even military personnel who admitted to having been part of a cover-up. Was the debris a weather balloon or something much more? Were alien bodies taken away for examination? Was there a survivor? The physical evidence collected was reportedly shipped out of Roswell to destinations unknown. What's more, many of the Roswell records appeared to have been destroyed.

In the years since, the U.S. Air Force changed its story several more times. In 1994, the military released a follow-up report on the incident. It said that the balloon testing was part of a secret program to monitor radiation levels and Soviet atomic activities. In 1997, the air force issued yet another paper, which argued that people's reports of UFO crashes and alien corpses had reasonable explanations. The bodies were actually human servicemen and military crash test dummies from accidents in the desert. The "alien" evidence was the result of the mistaken memories of people, collected over many years. Today people still debate what really happened near Roswell, New Mexico.

Contactees

The woman standing beside the reporter is brown-haired and well groomed, her bow-shaped mouth carefully painted with lipstick. In a cultured voice she tells the reporter, "I have made telepathic contact for the past eight years with many space people from many areas of space, both inside and outside this solar system." She is a "contactee," one of the many people who came forward in the 1950s claiming to have been contacted by space people.

J. Allen Hynek coined the phrase "close encounter of the first kind" to describe a UFO sighting at close range.

In the aftermath of World War II, everything was changing. The world was seeing the aftereffects of atom bombs dropped on the Japanese cities of Hiroshima and Nagasaki; the space age was dawning; and the Cold War between the United States and the Soviet Union was on everyone's mind. It was a time when anything seemed possible. The stage was set for people to treat reports of close encounters seriously.

In 1952, when George W. Van Tassel told the world that he had been in psychic contact with "Lutbunn, senior in command first wave, planet patrol, realms of Schare," it seemed too fantastic to believe, but people listened. As many as five thousand men and women flocked to the Giant Rock Interplanetary Spacecraft Conventions that Van Tassel founded, held in the California desert. These conventions continued through 1977 and included lectures, panel discussions, book and paraphernalia dealers, and contactees of every stripe telling their tales.

Another prominent contactee was George Adamski, a Polish immigrant to the United States. Adamski claimed, beginning in the late 1940s, to be in regular contact with people from Mars, Venus, and Saturn. He referred to these people as Space Brothers. Described as tall, remarkably humanlike, and very good looking, the Space Brothers had only one message: humanity should abandon its warlike ways, stop the coming nuclear holocaust, and enter an era of peace and abundance called the Cosmic Age.

Adamski wrote four books on the subject. His second, *Inside the Space Ships* (1959), detailed several meetings with the Space Brothers. Adamski wrote that sometimes he was compelled to drive into Los Angeles and check

CLOSE ENCOUNTERS DEFINED

In 1956, the National Investigative Committee on Aerial Phenomena (NICAP) was formed to study UFO encounters in an objective and scientific manner. NICAP investigated reports made by believable witnesses such as pilots, military officers, and scientists. The United States government investigated UFO reports under the code names Project Sign, Project Grudge, and Project Blue Book. These three projects seemed to concentrate on debunking the validity of UFO reports rather than trying to substantiate them. Soon, however, people began to claim that aliens from UFOs had abducted them.

Dr. J. Allen Hynek, who had worked on projects Sign, Grudge, and Blue Book, founded the Center for UFO Studies (CUFOS) in 1973. Dr. Hynek created the term "close encounters" as a way to classify the kinds of sightings and contact that people reported. He divided close encounters into the following categories:

- Close encounters of the first kind are UFO sightings within at least 150 yards (137 meters).
- Close encounters of the second kind are characterized by physical evidence such as burn marks on the ground or unidentifiable materials left at the location of a sighting.
- Close encounters of the third kind feature UFOs with visible occupants.
- Close encounters of the fourth kind are personal encounters with alien entities or abductions by aliens.
- Close encounters of the fifth kind consist of actual communication between a human and an alien.

into a hotel. Once there, he would go to the bar and be greeted by tall, handsome people who would escort him to their small, saucer-shaped scout ship, which was hidden in the desert. They would then fly off to the mother ship and take him to see the far side of the Moon, where thriving cities shined in

George Van Tassel said he designed the Integratron near Joshua Tree, California, in a joint effort with aliens. Intended as a rejuvenation and time machine, it is now used as a sound chamber.

the darkness. They showed him pictures of Venus, where metropolitan areas curiously like Los Angeles were located and where people lounged along the shores of crystal clear lakes. As proof of his contacts with the Space Brothers, Adamski presented photographs and home movies of the ships he claimed to encounter and diagrams of the ship interiors.

Expert examination, however, showed the photos and movies to be hoaxes, models suspended on fishing wire. Though Adamski's message of peace is still timely, the weight of evidence against him is overwhelming. With a surface shrouded by clouds that move 300 times faster than hurricane winds and an average surface temperature of 900 degrees Fahrenheit (482 degrees Celsius), Venus has no lakes. With no atmosphere or water source, the Moon harbors no sparkling cities on its dark side.

Eduard Meier, Truth Officer

In the 1970s, another contactee came to prominence. Eduard "Billy" Meier, a Swiss farmer, reported being contacted telepathically by visitors from the Pleiades, a cluster of stars. These beings traveled in crafts called beamships. The visitors said he was selected to be a "truth officer" to learn what they had to teach and told him to prepare for a difficult life because he would not be believed. Ultimately, Meier claimed, he developed a relationship with a beautiful blond Pleiadian named Semjase. He kept copious notes of their conversations and presented physical evidence he said was of Pleiadian origin. Meier also displayed remarkable photographs of the beamships. They were crisp, clear

One of the most famous "contactees" of the 1950s and 1960s was George Adamski, who claimed he encountered extraterrestrials in the California desert. Experts pronounced his photographs to be hoaxes.

images of flying saucers soaring over sunny Swiss mountainsides. There was only one problem: it was all a little too perfect.

Experts were never allowed to look at the original photographs or the negatives. Evidence, such as a metal fragment purported to be part of a beamship, somehow disappeared. Years later, after their divorce, Meier's former wife Kalliope denied that any of the story was true.

It's doubtful that any contactee actually had a close encounter of the fifth kind. Their stories left a legacy that would become a mixed blessing to the study of UFOs: a dispute between skeptics made wary by bizarre contactee tales and the open-minded, who would be astonished at the next generation of alien abduction accounts.

CHAPTER 2
The First Abduction Case

Betty and Barney Hill, an upstanding New England couple, reported the first major case of alien abduction in America. The story and evidence provided by Betty, a social worker, and Barney, a postal worker, changed the history of UFOs forever.

A Strange Trip

On the night of September 19, 1961, the Hills were driving south through the White Mountains in New England. It was a crisp, clear night with a full moon—perfect for seeing the stars that swept thick and bright over the New Hampshire landscape.

Betty, enjoying the view, suddenly noticed a very bright star. As the star seemed to get larger, she pointed it out to Barney.

BETTY HILL'S DREAM

Betty Hill's nightmare begins in her car on a deserted road somewhere in the New Hampshire night. There is a roadblock just ahead. A group of uniformed men walks toward the car. Betty and her husband don't know what they want, but they are coming closer and closer. Suddenly, Betty and Barney are no longer in the car. The uniformed men escort them into a strange, flat craft.

Betty watches helplessly as Barney disappears in the opposite direction, down a curved corridor and out of sight. The uniformed men tell her not to worry, that he'll be fine.

They take her to a room and tell her to undress. They give her a physical examination. She doesn't know what they want or why they're examining her. They put a needle in her navel and tell her it won't hurt, but it does and she begs them to stop. When one of them covers her eyes and tells her she's all right, the pain goes away. They assure her that no harm will come to her and that when it's all over, she'll be allowed to go home.

The light appeared to be following them. They stopped along the side of the road several times to try to figure out what they were seeing. Through their binoculars, Betty and Barney saw a cigar-shaped craft, a kind of airplane without wings. Around its edge, it had lights blinking in green, blue, red, and yellow. No aircraft moves that way, Betty insisted, but Barney wanted to believe it was something he could understand: a small, strangely quiet airplane. The truth was that Barney was afraid. He didn't know why, but there was something wrong with that light. Stars didn't move that way; satellites moved in only one direction; and regular airplanes couldn't maneuver in such a silent, jerky, mechanical fashion.

When they stopped again, Barney got out of the car and started walking toward the light. Through the binoculars, he saw a pancake-shaped ship with a double row of windows around its edge. In the windows he saw small figures moving back and forth. One of them was staring right at him. Betty yelled at him to come back to the car. Terrified, he ran back and jumped into the car, and they sped away. Suddenly he and Betty heard an electronic beeping coming from behind. The car shuddered. They felt a strange tingling, a sudden drowsiness, and then, nothing.

When the Hills arrived home, it was just after 5:00 AM, two hours later than they had expected. They felt clammy and nervous and were unable to remember parts of their drive. They both remembered the beeping sound. They also remembered a point in the trip when they were suddenly 35 miles (56 km) farther along the road than they'd been a moment before. What had happened in between, however, was a mystery.

Standing at the original site, Betty Hill describes how she and her husband, Barney, were abducted by aliens and taken aboard their craft.

Betty reported their sighting to the Air Police at nearby Pease Air Force Base. She wrote letters to UFO experts at NICAP. She cajoled Barney into returning to the places where they'd seen the craft, no matter how uncomfortable they both found doing so to be. Ten days after the sighting, Betty had a strange nightmare five nights in a row and then never again. There was also still the matter of the missing time. Why had it taken them two extra hours to get home?

Hypnotherapy Sessions

Two years passed. Barney developed ulcers and anxiety so acute that it interfered with his work. Finally, the Hills conferred with Dr. Benjamin Simon, a Boston psychiatrist who used hypnotherapy in his practice. In the doctor's office, under hypnosis, the two hours of missing time were revealed.

In chilling detail, Barney recalled a strange group of men stopping him and his wife along the road. In particular, he remembered the leader's eyes. Barney cried out, "His eyes were slanted. Oh—his eyes were slanted!" He added, "I've never seen eyes like that before." Barney described how the men escorted them through the woods to the ship he'd seen in the sky. He and Betty were taken to separate rooms inside the ship. There, he underwent an intimate and upsetting physical examination. When the exam was finished, he and Betty were returned to the car and sent on their way. Dr. Simon recorded all of Barney's hypnotic sessions. He planted a suggestion that Barney wouldn't remember the sessions until he was prompted. Dr. Simon then hypnotized Betty to see what she remembered.

Detailed memories of an abduction emerged during the Hills' hypnotherapy sessions with Dr. Benjamin Simon in the 1960s.

Betty's tale was even more bizarre. Just as she had dreamed, she remembered being taken aboard a craft hidden in the woods by the roadside. She described a physical examination that included the sampling of hair, skin, and fingernails, along with the painful needle inserted into her navel. She recalled the alien beings telling her this was a pregnancy test. She said that after the

Betty Hill claimed that one of the aliens showed her a star map. Her reproduction of the map proved to be scientifically accurate.

examination, she talked with one of the beings and that he showed her a book and a star chart. Finally, she said, she and Barney were returned to the car and watched the ship depart.

In the weeks that followed, Dr. Simon worked with the Hills, testing their stories to see if any details changed. He was trying to determine if Betty had planted the abduction idea in Barney's mind, if the story was a hoax, or if it was the result of a psychotic episode. Neither the doctor nor the government experts were ever able to prove anything.

The image of the star map that Betty recalled stayed with her, and she drew a reproduction. It detailed not only stars important to the visitors, but also what were described as trade routes and regular corridors of travel. On April 13, 1965, the *New York Times* published a star chart as part of an article on a new radio source in space called CTA-102. The similarities between the chart published in the paper and the patterns in Betty's star map were striking.

The peculiar pregnancy test was the item that intrigued investigators most at the time. In the early 1960s, there were no invasive pregnancy tests being conducted. Further, it wasn't until years later that laparoscopy—the practice of inserting a long, thin tube fitted with fiber optics into a patient's navel for internal observation and surgical procedures—would be developed and used as a regular technique. How Betty Hill could predict such a development in medical technology remains a mystery even today.

CHAPTER 3

Other Famous Encounters

In the years since Betty and Barney Hill's experience, thousands of other people have reported close encounters with extraterrestrial beings. These people have told stories of extraordinary experiences, including abduction by aliens. Whether or not these tales have had anything to do with reality has not been clear. This chapter explores two of the most famous cases in the history of alien encounters.

The Kathie Davis Case

By the time that Kathie Davis (a pseudonym, or false name) wrote to Budd Hopkins, a leading investigator in the field of alien abduction, her life was filled with signs of strangeness. She

Kathie Davis and other abductees have described alien beings they call "Grays." Typical features include gray skin; large, egg-shaped heads; big, black eyes; and small, slit-like mouths and nostrils.

had scars on her body that she couldn't explain. There were geometric patterns in the yard behind her home where grass had stopped growing. She had dreams of gray-faced beings, or "Grays," coming into her home at all hours of the night. Her sister, Laura, had experienced periods of missing time. Her mother had seen lights move in and out of the backyard.

Davis wrote to Hopkins after reading his book, *Missing Time*, which described his work. To her dismay, Davis realized that her personal experiences matched much of what she read. But there was one major difference: she wasn't the only person in her family to have the peculiar dreams, the strange marks on the body, and the creeping anxiety. She also had experiences that offered physical evidence.

In June 1983, Kathie's mother, Mary, saw a strange light moving through the backyard. Later that evening, Kathie and two friends went for a swim in the backyard pool and found themselves feeling nauseated. The next day, the ground in the yard appeared to have been burned; the Davis's dog wouldn't go anywhere near it. Over the course of the next three days, the grass in the burned area died off completely. Kathie and her friends suffered continuing nausea. Kathie lost some hair and suffered a sudden onset of eye irritation. In addition, small, scoop-shaped scars appeared on both her and her mother's knees. Neither knew what had caused them.

Hopkins recognized the classic signs of a possible close encounter: the scoop marks on the body, a free-floating anxiety that had no direct association to a clear memory, periods of missing time, and strange marks on the ground. Before he knew it, his list of people to interview included the entire Davis family and neighbors next door and across the street.

First, Hopkins examined the physical evidence. He had soil samples from the Davis yard tested. The burned soil was chemically identical to the normal soil, but it appeared to have been subjected to extremely high temperatures for a long period of time.

Next, Hopkins investigated Kathie's experiences. Under hypnosis, she recalled encounters and abductions that went as far back as childhood. In all of those experiences, the common element was the appearance of small, gray people who would meet her and take her elsewhere. She recalled that they had large heads, big black eyes, and small mouths. They seemed to speak to her telepathically. The most significant recollections were of her abductions in the late 1970s and early 1980s.

In late 1977, Kathie Davis was in a car with two friends when they sighted a UFO. Like Betty and Barney Hill, Kathie recalled herself being paralyzed and taken somewhere to be subjected to an unpleasant and unnerving physical examination. At the time, Kathie was engaged to be married. Shortly after the incident, to her delight, she discovered she was pregnant. She and her fiancé moved their wedding date so they could be married sooner. Kathie recalled that just a few months later, in March of 1978, she had another encounter that changed everything.

Kathie was staying at her sister and brother-in-law's home and was feeling nervous and unwell. She attributed this to her pregnancy, still in its early stages. She started to feel drowsy and then, though she was alone in the house, felt someone massaging her lower back and shoulders. At first, the experience was comforting, but it became frightening as she felt herself being opened up like a flower. First there was pain, then fear.

WHAT TO DO IF YOU SEE A UFO

A UFO is an object or light seen in the sky or on land, whose appearance, motions, lights, and colors do not have a logical or natural explanation. Founded in 1969, the Mutual UFO Network, Inc. (MUFON) is an organization dedicated to resolving the mystery of UFOs through the efforts of volunteers. People interested in UFO events participate as investigators, amateur astronomers, and journal contributors. On the MUFON Web site (http://www.mufon.com), an article by UFO enthusiast Michael Curta lists things you can do if you encounter an unidentified flying object. Some of his tips are:

- **Remain calm.** If you are calm, it will be easier to remember all of the details of the event.

- **Protect yourself from any hazards.** Be prepared to take any necessary action to get out of the way or hide.

- **Record the event.** Use a camcorder or camera to record what you are seeing. Try to keep reference points in the field of view. As an alternative, draw pictures of what you are seeing and the area around it. If you have the ability to record audio, record your description of the event as it is occurring.

- **Get the help of witnesses.** If other witnesses are present ask them to also write or record their own observations. Do not discuss the details with each other until everyone is finished.

- **Estimate.** Try to judge the distance from you to the object, the object's altitude, and its speed. For example, did the UFO cross the sky in five seconds or five minutes?

- **Be objective.** Eliminate every other possibility before considering the possibility that you have seen an actual UFO.

If you would like to share what you have seen with others, you can report the event to a UFO research organization or share the sighting on a Web site for UFO watchers.

In the next days, Kathie discovered she was no longer pregnant. She hadn't experienced any of the usual signs of a miscarriage, but suddenly the baby she had been carrying was gone. Her physician couldn't explain it. To Kathie's thinking, it appeared that the gray beings had taken her baby.

In 1983, Kathie recalled, she had yet another abduction experience. She remembered standing in a room with four Grays. She felt anticipation growing in the room. A door opened, and two more Grays came in with a little girl with wispy white hair and large blue eyes. She had a small pink mouth, a tiny nose, and a larger-than-normal head for a child her size. One of the Grays told Kathie to be proud, that this little girl was part of her, but that the little girl must stay with them. Kathie was certain that this was the child the Grays had taken from her. They promised Kathie that she would see the little girl again in the future and then sent her on her way.

Budd Hopkins met a number of other women besides Kathie Davis who told similar stories. These women also spoke about having pregnancies that ended prematurely and later being presented with strangely delicate, big-headed children. Other researchers have heard similar stories told by women who were completely unfamiliar with Hopkins's work or the Davis story. Such stories continue to be told to this day by individuals who otherwise seem completely normal.

The Travis Walton Case

The abduction case of Travis Walton is another landmark in the history of alien investigation. In November 1975, Walton and six coworkers were in the

The film *Fire in the Sky* dramatizes Travis Walton's abduction experience in 1975. In this scene, Walton gazes up at the light beam that reportedly knocked him over before his disappearance.

mountainous Apache-Sitgreaves National Forest in eastern Arizona. As loggers, their job was to thin a section of the forest to allow for faster growth. One evening, as they slowly drove their pickup truck down the winding mountain road, they noticed a bright light in the woods. At first they thought this was the light from a crashed airplane, but as they drove closer they realized that what they were seeing was no ordinary aircraft. It was a golden, glowing saucer hovering 90 feet (27 m) off the ground.

Walton, the daredevil of the group, got out of the truck and approached the saucer to get a better look. His friends urged him to come back, but he refused to listen. As he got closer, Walton heard a quiet mechanical whine. Suddenly, the whine grew into an overwhelming roar. The saucer began to wobble. Walton started back to the truck. A dazzling blue-green beam shot out of the underside of the saucer and hit Walton in the chest, lifting him off the ground and throwing him aside about 10 feet (3 m). His body landed on the ground and he lay still. Mike Rogers, the driver, panicked. He threw the truck into gear and sped away.

Half a mile down the road, Rogers stopped the truck, realizing that he'd left his best friend behind. Although all the men were frightened by what had just happened, they couldn't in good conscience leave Walton stranded and possibly injured. They drove back for him. When they returned to the spot, however, Travis Walton was gone. So was the flying saucer.

The following days were a flurry of activity as the news of Walton's disappearance spread. The local police organized search parties to comb the mountain forests. UFO researchers from a group called Ground Saucer

Watch came to take soil samples and readings for radiation, magnetism, and ozone. The police questioned Rogers and the others closely, suspecting murder, kidnapping, or fraud of some kind. Five of the six men passed lie detector tests, clearing them of any crime and confirming their belief that they had seen a flying saucer. Later testing confirmed the truthfulness of the sixth man as well. Suspicion remained. Five days later, Travis Walton suddenly returned.

Walton returned to consciousness on the side of a mountain road and, in a daze, called family to pick him up. He thought he'd been gone for only two hours. His five-day beard growth told a different story. Over the next weeks, Walton's family tried to shield him from the press while he recovered. The Aerial Phenomena Research Organization (APRO) and reporters from the *National Enquirer* befriended Walton. These organizations sponsored much of the treatment Walton underwent in the wake of his return. He had a physical, had blood and urine samples taken, underwent hypnosis, and took a lie detector test, which he failed due to stress. Walton later passed a lie detector test with more than a 90 percent certainty of truthfulness.

The Walton case is perhaps the best-documented abduction case in the history of ufology. Seven people all say they witnessed the same thing. Trace evidence from the area in question testifies that something unusual happened there; the evidence is consistent with the findings from other UFO sighting locations. One skeptic, an investigator named Jerry Black, after delving as deeply as he could into the case, couldn't punch any holes in the story. The case is unique in almost every way.

In a court of law, the testimony of seven witnesses along with physical evidence and records from experts in several different disciplines would typically create a compelling case for any lawyer—except when the case is one of alien abduction. Despite all the supporting evidence, the nature of so fantastic and unnerving a subject makes its possibility hard to accept. Skeptics are still challenging the story almost thirty years later. They put forward the theory known as Occam's razor, which states that, all things being equal, the simplest explanation tends to be the correct one. Perhaps the Travis Walton case is the exception that proves the rule. The simplest explanation may be the correct one, but once the simplest explanations are eliminated, the fantastic suddenly becomes possible.

Chapter 4

Explaining Close Encounters

Talk to a believer or a UFO investigator and the message is clear: Something otherworldly is happening on Earth. Dr. David M. Jacobs, a history professor at Temple University and the author of *Secret Life: Firsthand Documented Accounts of UFO Abductions*, says, "Even if there is only the smallest percentage of a chance that [the alien abduction phenomenon] is real, we should begin to put energy and funding into studying it because the payoff is so enormous. It demands serious attention."

At the other end of the spectrum is Philip J. Klass, the most well-known UFO skeptic. Klass is a retired senior avionics editor of *Aviation Week & Space Technology* magazine and one of the founders of the Committee for the Scientific Investigation

A car on the property of the Unarius Academy of Science encourages people to welcome alien visitors. The organization purchased a site in California where "space brothers" can land.

of Claims of the Paranormal (CSICOP). He's on record for offering $10,000 to the first person who reports his or her UFO abduction to the Federal Bureau of Investigation (FBI), the law enforcement agency responsible for kidnapping, and whose report the FBI believes. Klass told PBS's *Nova*, "If extraterrestrials are abducting earthlings, as is claimed, then it is time to alert the federal government to defend us, for our government to join with other governments to defend this planet...And if this is simply fantasy, then let's dispel it, let's push it off our plate of things to worry about."

Except for their stated interest in discovering the truth—whatever that may be—Jacobs and Klass have little in common. In fact, the controversy over the possibility of alien contact and abduction has fostered heated conflict in the press, in books, and at conferences across the United States.

Part of the debate focuses on the reason for such abductions: Even if we believed such kidnappings were possible, why would visitors from another world be interested in human beings? What could be of interest to beings advanced enough to master star flight? One might ask the question in another context: why would humans be interested in dolphins, pandas, or Siberian tigers? Catch-and-release programs for exotic species have been part of regular scientific practice in recent decades. Could alien motivation be so different from our own when the abduction scenario suggests such a familiar study method?

A Psychological Explanation

The stories of those who believe they have been abducted by aliens often have striking similarities. People remember feeling paralyzed and extremely frightened. They report seeing flashing lights, hearing loud sounds, sensing the presence of horrifying intruders, and feeling probing and pain. Some psychologists say that these similarities are not necessarily evidence that the experiences were real. At the same time, psychologists do not believe that the abductees are necessarily all crazy or mentally ill.

Dr. Susan A. Clancy, a Harvard-trained psychologist, links these experiences to a condition called sleep paralysis. In this condition, a person can wake up

ARE ALIEN SIGHTINGS A SCIENTIFIC EFFECT?

Another theory being proposed today is that alien encounters are nothing more than encounters with our own biology and the geology of the earth. In Canada, researchers are exploring the connection between geological activity, UFO sightings, and reports of alien abduction. When tectonic plates move, rubbing against each other and breaking and changing shape, energy is released (like striking a flint to produce a spark). The theory goes that spots of light described as UFOs are, in fact, energy flashes resulting from plate tectonics. Such released energy might stimulate the brain, specifically the temporal lobe, to produce images and experiences in sleep or unconsciousness that resemble the experiences that abductees describe under hypnosis. Experiments at Laurentian University in Ontario have demonstrated that sensations such as paralysis, the feeling of others in a room, fear and paranoia, and visions of gray, waxy-faced beings can be provoked by stimulating parts of the brain with minor magnetic waves.

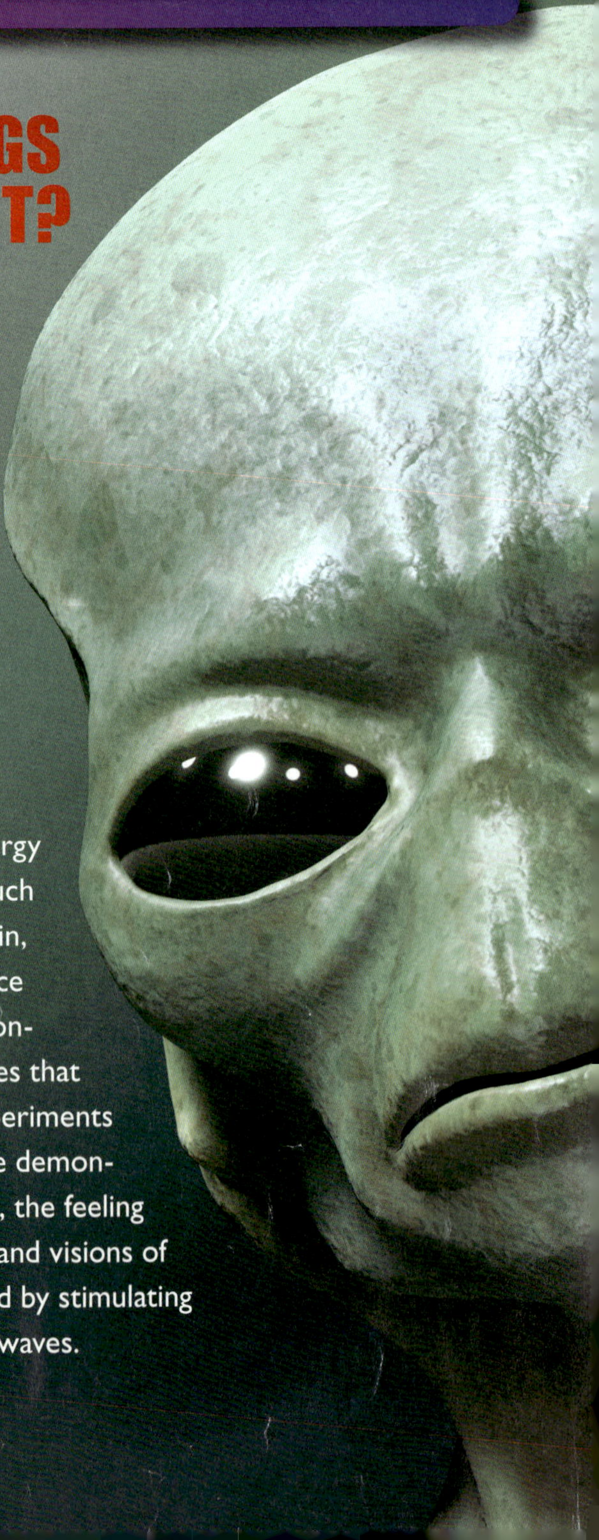

while still paralyzed in a dream state. In a 2005 interview with the *Harvard Gazette*, Dr. Clancy said, "We can find ourselves hallucinating sights, sounds, and bodily sensations. They seem real, but they're actually the product of our imagination." While she believes that these sensations arise from anxiety and a vivid fantasy life, some people may feel the need to find a different explanation. According to Dr. Clancy, those who hold beliefs in the paranormal may find that blaming an alien encounter is a way to explain their pain and distress.

According to the tectonic stress theory, stress within Earth's crust in an earthquake zone, such as the San Andreas Fault, can give rise to lights in the nearby sky.

Clancy and other social scientists report that people usually come to these conclusions gradually, over a period of time. In the book *Extreme Deviance*, sociologist Christopher Bader explains that believers have often found their way to other people who support their interpretation of events. For example, they might visit a therapist with an interest in UFOs or attend meetings of a UFO support group. These interactions strengthen their belief that they were truly abducted.

However, Dr. Jacobs disagrees that the scientific case is closed on alien abductions. In a 2006 article in the *Journal of Scientific Exploration*, Jacobs reminds readers, "During abduction events, abductees are missing from their normal environments. Police have been called, search parties have been sent out, parents have frantically searched for their children, etc…People are abducted while fully awake, driving a car, gardening, and so forth." Jacobs states that according to his research, less than half of all abductions take place at night while people are in bed. He argues that one cannot explain away all close encounters as nightmares and false memories.

Cultural Influences

Another part of the UFO debate focuses on the idea of cultural contamination. So many of us have seen movies like *Close Encounters of the Third Kind* or watched television shows like *The X-Files* that, skeptics say, it's hard to find someone who hasn't been exposed to the idea of abduction or the image of a gray-skinned alien with a big head and large, almond-shaped eyes. Such images, they say, have so permeated our culture that even people who honestly believe they've been abducted can't be considered reliable: they may have dreamed their experiences and unconsciously incorporated popular imagery into their memories. However, many investigators, like Hopkins and Jacobs, say they always withhold certain details from their publicized cases and keep them confidential. These details, they say, are consistent from case to case, are not found in the movies or on TV, and allow them to recognize a genuine abduction experience.

A Long Tradition

Investigators are now examining UFO sightings and alien encounters though the eyes of history, religion, folklore, and sociology. After all, from generation to generation, stories of encounters with strange, otherworldly crafts and creatures survive. Has this phenomenon, then, always been with us?

The earliest answer might be found in some of our oldest literature. The Bible includes stories like Ezekiel's vision: Ezekiel describes four creatures, each with four faces, coming out of the sky in a flash of light and fire and wind, riding on a "wheel within a wheel," which suggests something disc-shaped. Ezekiel Chapter 1 contains the following verses:

> *And I looked, and, behold, a strong wind came out of the north, a great cloud, with a fire flashing up, so that a brightness was round about it... And out of the midst thereof came the likeness of four living creatures. And this was their appearance: they had the likeness of a man, but every one of them had four faces, and every one of them had four wings...As for the*

Stories of strange objects in the sky and other-worldly visitors have a long history. In the Bible, Ezekiel has a vision of four mystical beings traveling on wheel-like objects.

likeness of their faces, each had the face of a man in front; the four had the face of a lion on the right side; and the four had the face of an ox on the left side; and the four had the face of an eagle at the back... Now as I looked at the living creatures, I saw a wheel upon the earth beside the living creatures, one for each of the four of them...The appearance of the wheels and their construction was like unto the color of beryl...and their construction was as it were a wheel within a wheel.

People in the Middle Ages described being visited in the night by incubi (male) and succubi (female), demons that would slip into their victims' rooms with the purpose of sinful seduction. There are many European and Asian folktales about magical beings abducting people. The Irish tell tales of leprechauns who lure people away with the bait of hidden treasures.

Even as late as the 1890s, with the Industrial Age in full swing and new scientific marvels being invented every year, people still told stories of seeing things in the sky. From Chicago to San Francisco, people claimed to have observed cigar-shaped ships or dirigibles flying with lights shining down to the ground. Such stories were reported in newspapers, and some even suggested that men from the Moon were building airships.

Every society tries to explain mysteries in terms of its own frame of reference. In our space age, the flying saucers move at rocket speed. During World War II, pilots reported seeing "foo fighters," fast-moving, glowing objects following their aircraft. The stories from the 1890s, before the age of airplanes, describe the ships as dirigibles. In the ancient eras, people did not contemplate aliens flying aircraft from other planets. They did, however, speculate about creatures from the Otherworld sent by the gods. What might have appeared to a nun in medieval England as an incubus might be described today as an alien with an interest in creating an alien-human hybrid. Perhaps yesterday's fairies and demons are today's alien visitors in a different guise.

Chapter 5

Aliens in Popular Culture

After the Kenneth Arnold and Roswell sightings in 1947, there was a frenzy of popular interest in UFOs and aliens. The phenomenon was soon taken up by Hollywood, which turned out numerous movies featuring otherworldly creatures in the following years. The popularity of these films, as well as books, television shows, and comics, showed that even people who did not believe in UFOs could be entertained and thrilled by the idea of alien visitations.

Alien Films of the 1950s

The early 1950s were a rich time for movies about UFOs and aliens. One of the first examples of the genre was the

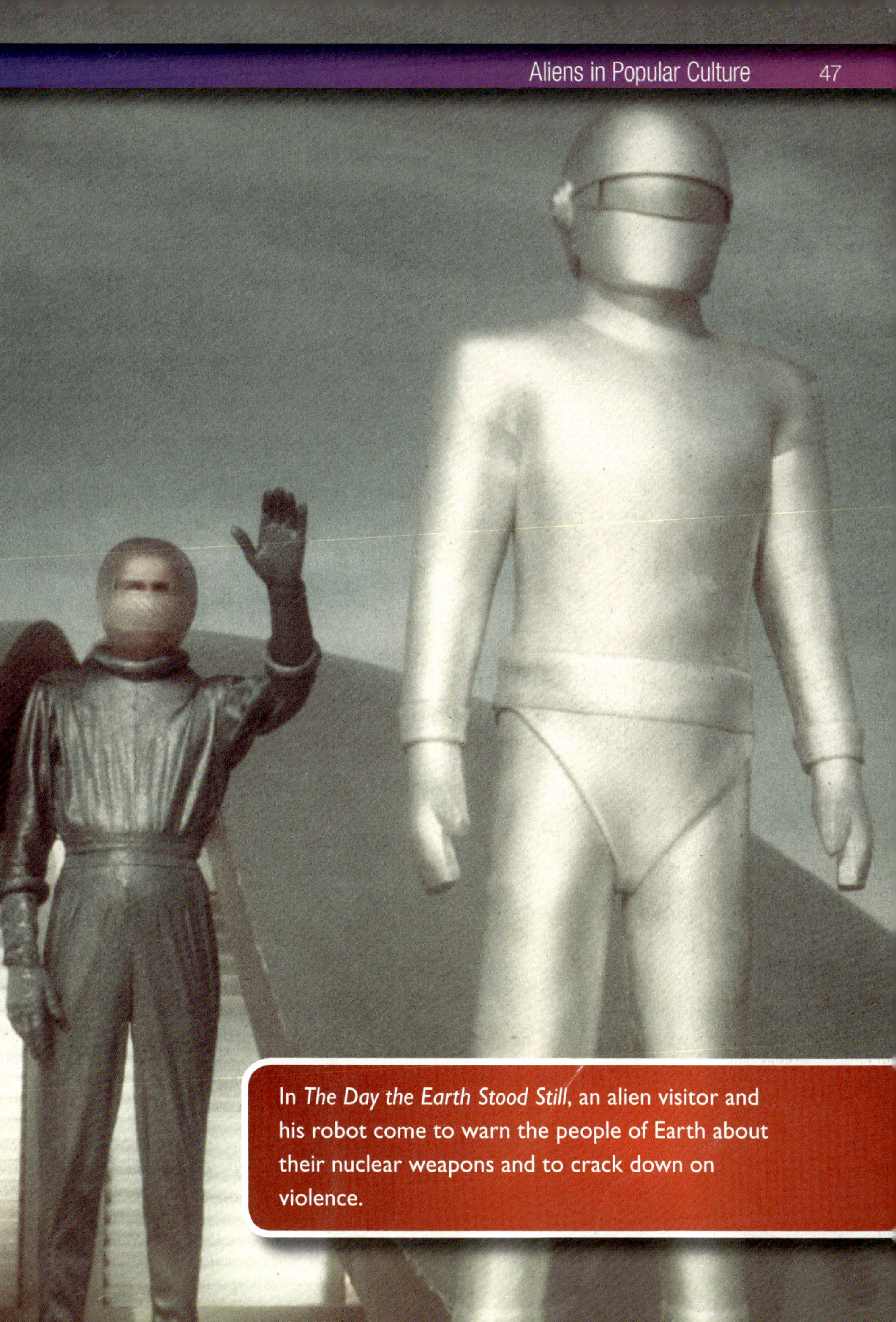

In *The Day the Earth Stood Still*, an alien visitor and his robot come to warn the people of Earth about their nuclear weapons and to crack down on violence.

1951 film *The Day the Earth Stood Still*. In this film, an alien named Klaatu arrives in Washington, D.C., in a flying saucer. Although he announces that he and his companions have come in peace and goodwill, the arrival triggers panic and chaos. While U.S. officials race to deal with the problem, a widow and her young son become involved in Klaatu's mission. A nervous soldier shoots Klaatu, but he comes back to life and delivers a message from the Galactic Federation: the development of atomic weapons on Earth will not be permitted to get out of hand. Either Earth must live in peace under the rule of Klaatu's robot servant Gort or it will be destroyed. The movie's themes reflected Americans' fascination with aliens, as well as their political concerns and fears during the Cold War. A remake of the film starring Keanu Reeves, Jennifer Connelly, and Kathy Bates was released in 2008.

Following this film, a rash of other movies came out featuring alien invasions. Kids and adults rushed to the movies in droves to be pleasurably scared out of their wits by visions of interplanetary monsters attacking the earth. These films included *The Thing* (1951), *Invaders from Mars* (1953), *The War of the Worlds* (1953), *Invasion of the Body Snatchers* (1956), and *Earth Versus the Flying Saucers* (1956).

In addition to entertaining popular audiences, these films had a strong influence on the growing ufology movement. UFO skeptics have noted that the stories of contactees such as George Adamski and Betty and Barney Hill often had details that strongly resembled the images portrayed in the movies.

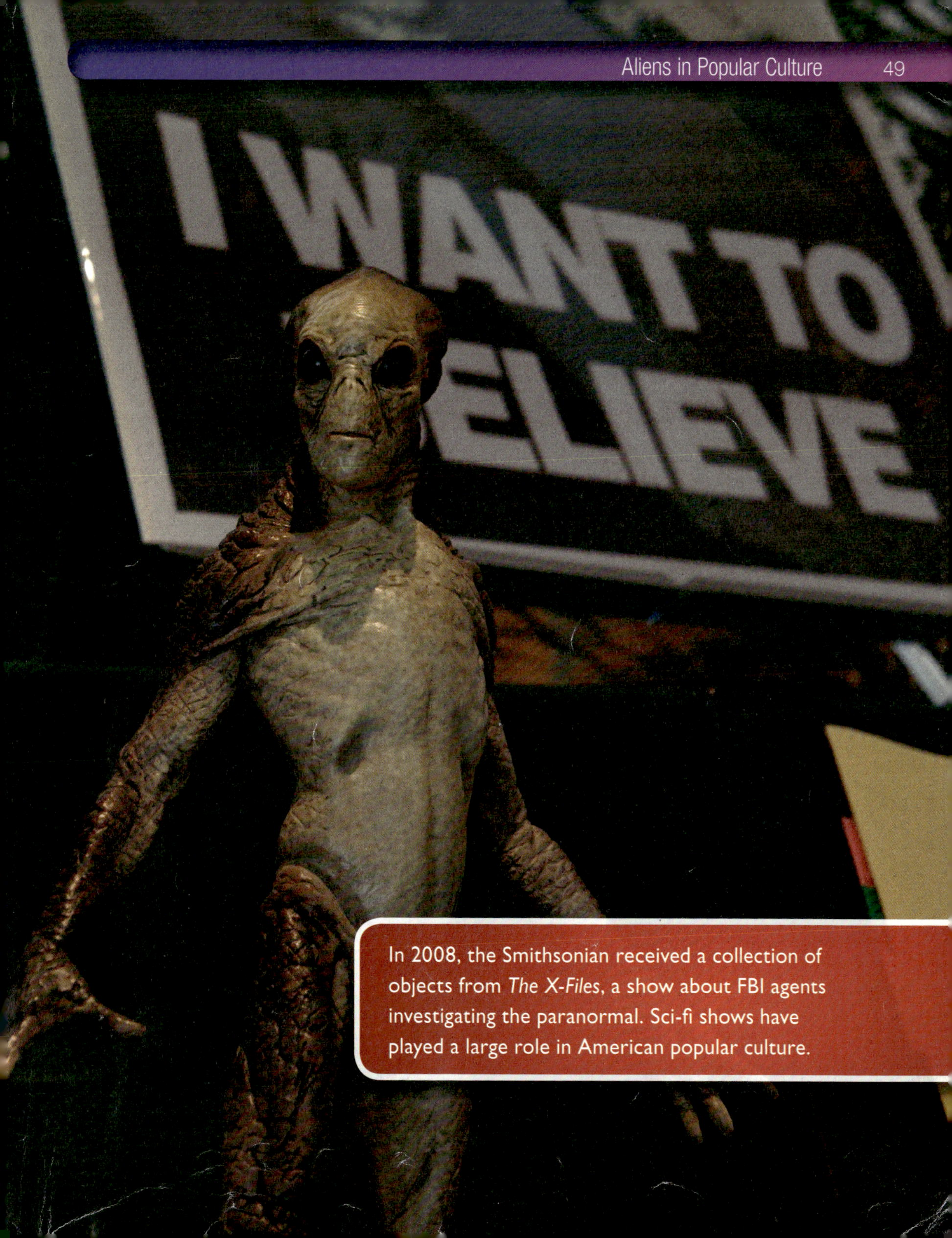

In 2008, the Smithsonian received a collection of objects from *The X-Files*, a show about FBI agents investigating the paranormal. Sci-fi shows have played a large role in American popular culture.

Another Dimension: 1960s Television

In the 1960s, science fiction began to occupy more time on television. Groundbreaking shows like *The Twilight Zone*, created by Rod Serling, began to draw a wide audience. *The Twilight Zone* was an anthology that used different actors and plotlines for each episode. The show dramatized riveting science fiction stories by authors such as Ray Bradbury, Richard Matheson, and Charles Beaumont. A similar anthology show called *The Outer Limits* also aired during the early 1960s. Episodes of both shows featured creepy scenarios, including many instances of aliens invading and colonizing the earth. Alien abductions and alien possessions were also common themes.

In turn, these shows influenced the most famous sci-fi TV show of all—*Star Trek*. Created by Gene Roddenberry, the show hit the television airwaves in 1966. Set in the future, the show presented the exploration of space as an exciting new frontier. *Star Trek* followed the adventures of the crew of the U.S.S. *Enterprise*, led by Captain James T. Kirk and Mr. Spock. Part of an interplanetary alliance known as the Federation, the *Enterprise*'s mission was "to seek out new life and new civilizations, to boldly go where no man has gone before." Episodes of the show brought the crew into contact with new alien races or face-to-face with the menacing Klingons and Romulans. Although the first version of the series ran for only three years, *Star Trek* became a cult classic, generating intense fan groups. In recent years, movies and new generations of the TV series have been set in this fictional world, including *Star Trek: The Next Generation*, *Star Trek: Deep Space Nine*, and *Star Trek: Voyager*.

LOL ALIENS!

In addition to fascinating and frightening audiences, some television shows and films have explored the lighter side of alien life. Many of these productions have mined the comic potential of landing characters on Earth who are unfamiliar with human ways. Some of the many TV shows and movies that have made audiences laugh out loud are:

- *My Favorite Martian* (1963). A man named Tim helps a stranded Martian get along on Earth by passing him off as Uncle Martin.

- *Lost in Space* (1965). The Robinson family was going into space to fight for a chance for humanity when a spy sent to sabotage the mission throws their ship off course and becomes trapped with them.

- *Mork & Mindy* (1978). In a spinoff from the popular TV show *Happy Days*, Robin Williams plays the wacky alien Mork from Ork, who falls in love with and marries the earthling Mindy.

- *3rd Rock from the Sun* (1996). A group of aliens has come to Earth to learn about people and their customs. To avoid detection, they have taken on human form: the leader masquerades as a college professor and the crew as his family.

- *Men in Black* (1997). This movie follows the adventures of Kay and Jay, members of a top-secret organization established to monitor and deal with alien activity on Earth. *Men in Black III* comes out in 2012.

- *Galaxy Quest* (1999). In this film, a group of aliens approach actors from an old science fiction series to help them with an intergalactic rescue mission. At first, the actors assume the aliens are just fans who like to dress up and attend sci-fi conventions. However, the actors soon realize they are working with real extraterrestrials.

- *Monsters vs. Aliens* (2009). After a space meteor hits a young bride-to-be and turns her into a giant monster, the military captures her and takes her to a secret government compound. There, she meets a number of other monsters in detention. They are all called upon to save the Earth when a mysterious alien robot attacks.

Steven Spielberg's Megahits

Director Steven Spielberg has made some of the most creative, popular, and highest-earning films of all time. Two of his biggest hits of the 1970s and 1980s focused on the subject of alien-human encounters.

In *Close Encounters of the Third Kind* (1977), Roy Neary, an electrical lineman, is going to investigate a power outage when his truck stalls. He witnesses the lights of a UFO in the night sky and even sustains a sunburn from them. After this experience, strange mountain-like images and five musical notes keep running through his mind. Roy becomes obsessed with learning what these signs represent, refusing to accept a logical explanation. He leaves home to attempt to learn the truth about UFOs. Meanwhile, other strange events are taking place across the world. Government agents and UFO experts, including a French researcher who believes people can use music to communicate with aliens, are investigating. Eventually, everyone's theories and beliefs are put to the test in an isolated area in the wilderness.

In *E.T.: The Extraterrestrial* (1982), a group of alien botanists is studying plants on Earth. When a human task force discovers them, the aliens take off quickly and leave one of their members behind. The lovable little alien finds himself isolated on a strange planet and unsure of how to get home. Fortunately, the extraterrestrial soon meets Elliott, a boy who becomes his friend and names him "E.T." Elliott hides E.T. and learns to communicate with him telepathically. E.T. wants desperately to get home, and Elliott tries to help

At the end of the movie *Close Encounters of the Third Kind*, the aliens' mother ship flies off peacefully into the night sky.

him. However, he must survive testing by government scientists and serious sickness before he can be rescued. In 2002, Steven Spielberg released a revised edition of the movie, which warmed millions of hearts and once again had everyone repeating the phrase "E.T. phone home."

Contemplating the possibility of intelligent life far beyond the Earth is a topic that has fascinated many scientists, authors, artists, and everyday people. Whether or not such beings truly exist and can visit us from places far away, people will continue to enjoy gazing into the night sky and asking, "Could it be?"

GLOSSARY

abductee A person who has been abducted.

abduction The illegal kidnapping of a person. In ufology, abduction refers to the capture, examination, and return of human beings by alien visitors.

alien A being from another planet; extraterrestrial.

close encounter Sighting of or contact with a UFO or aliens. There are five categories of encounters, ranging from sightings to actual communication between a human and an alien.

contactee A person who claims to have had contact with aliens.

flying saucer Popular term used to describe an oblong-shaped unidentified flying object.

hypnosis A trancelike state of relaxation and concentration, often characterized by vivid recall of memories and fantasies.

hypnotherapy Treatment of an emotional problem, disease, or addiction by means of hypnosis.

skeptic A person who questions or doubts accepted beliefs.

ufology The study of UFOs and related phenomena.

unidentified flying object (UFO) An object that flies through the air but bears no resemblance in shape or behavior to any recognizable form of conventional aircraft.

FOR MORE INFORMATION

Canadian Space Agency
John H. Chapman Space Centre
6767 Route de l'Aéroport
Saint-Hubert, QB J3Y 8Y9
Canada
(450) 926-4800
Web site: http://www.asc-csa.gc.ca
The Canadian Space Agency is committed to leading the development and application of space knowledge for the benefit of Canadians and humanity.

Committee for Skeptical Inquiry
Box 703
Amherst, NY 14226
(716) 636-1425
Web site: http://www.csicop.org
The mission of the Committee for Skeptical Inquiry is to promote scientific inquiry, critical investigation, and the use of reason in examining controversial and extraordinary claims.

International UFO Museum and Research Center
114 North Main Street
Roswell, NM 88203
(800) 822-3545
Web site: http://www.roswellufomuseum.com

People come to this local museum for more information about the 1947 Roswell incident and other UFO phenomena. The museum helps host the annual Roswell UFO Festival during the first week of July to celebrate the anniversary of the event.

The J. Allen Hynek Center for UFO Studies
P.O. Box 31335
Chicago, IL 60631
(773) 271-3611
Web site: http://www.cufos.org
The Center for UFO Studies (CUFOS) is an international group of scientists, academics, investigators, and volunteers dedicated to the continuing examination and analysis of the UFO phenomenon.

National Aeronautics and Space Administration (NASA)
Public Communications Office
Suite 5K39
Washington, DC 20546-0001
(202) 358-0001
Web site: http://www.nasa.gov
NASA's mission is to pioneer the future in space exploration, scientific discovery, and aeronautics research.

SETI Institute
189 Bernardo Avenue, Suite 100

Mountain View, CA 94043

(650) 961-6633

Web site: http://www.seti.org

The mission of the SETI Institute is to explore, understand, and explain the origin, nature, and prevalence of life in the universe.

Web Sites

Due to the changing nature of Internet links, Rosen Publishing has developed an online list of Web sites related to the subject of this book. This site is updated regularly. Please use this link to access the list:

http://www.rosenlinks.com/me/ali

FOR FURTHER READING

Burgan, Michael. *Searching for Aliens, UFOs, and Men in Black* (Unexplained Phenomena). Mankato, MN: Capstone Press, 2011.

Davis, Barbara J. *The Kids' Guide to Aliens* (Kids' Guides). Mankato, MN: Capstone Press, 2010.

DeMolay, Jack. *UFOs: The Roswell Incident* (Jr. Graphic Mysteries). New York, NY: PowerKids Press, 2007.

Duncan, John. *UFOs* (Unexplained). Milwaukee, WI: Gareth Stevens Publishing, 2006.

Grace, N.B. *UFO Mysteries* (Boys Rock!) Chanhassen, MN: Childs World, 2006.

Hamilton, John. *Aliens: The World of Science Fiction*. Edina, MN: ABDO Publishing Co., 2007.

Kallen, Stuart A. *Alien Abductions* (The Mysterious & Unknown). San Diego, CA: ReferencePoint Press, 2008.

Keene, Carolyn. *Close Encounters* (Nancy Drew: Girl Detective). New York, NY: Aladdin Paperbacks, 2007.

Klass, David. *Stuck on Earth*. New York, NY: Frances Foster Books, 2010.

Kneece, Mark, and Rod Serling. *The Twilight Zone: Will the Real Martian Please Stand Up?* New York, NY: Walker & Co., 2009.

Mason, Paul. *Investigating UFOs and Aliens* (Extreme!) Mankato, MN: Capstone Press, 2009.

Netzley, Patricia D. *Alien Encounters* (Extraterrestrial Life Series). San Diego, CA: ReferencePoint Press, 2012.

Parks, Peggy J. *Aliens* (Mysterious Encounters). Detroit, MI: KidHaven Press, 2007.

Rex, Adam. *The True Meaning of Smekday*. New York, NY: Hyperion Books for Children, 2007.

Rooney, Anne. *UFOs and Aliens* (Amazing Mysteries). Mankato, MN: Smart Apple Media, 2010.

Shostak, G. Seth. *Confessions of an Alien Hunter: A Scientist's Search for Extraterrestrial Intelligence*. Washington, DC: National Geographic, 2009.

Skurzynski, Gloria. *Are We Alone? Scientists Search for Life in Space*. Washington, DC: National Geographic Society, 2004.

Southwell, David, and Sean Twist. *Unsolved Extraterrestrial Mysteries* (Mysteries and Conspiracies). New York, NY: Rosen Publishing, 2008.

Teague, Mark. *The Doom Machine: A Novel*. New York, NY: Blue Sky Press, 2009.

INDEX

About the Authors

Ian M. Goldfarb is an author and educator in New York City.

Janna Silverstein is a writer, editor, and teacher who has published nonfiction in print and on the Web on a variety of subjects, including books, technology, and travel. Her short fiction has appeared in several anthologies as well as in *Marion Zimmer Bradley's Fantasy Magazine*. She has been watching the skies since her childhood, but to her disappointment, she has never seen a UFO or experienced a close encounter.

Photo Credits

Cover, back cover, p. 1 (alien) Zap Art/Taxi/Getty Images; cover, back cover, p. 1 (background), 6, 8, 12–13, 18, 19, 26, 30–31, 37, 40, 46, 51, 52 Shutterstock; cover, back cover, p. 1 (camera lens) © www.istockphoto.com/jsemeniuk; p. 5 © Columbia Pictures/Courtesy Everett Collection; p. 10 Ann Cutting/Workbook Stock/Getty Images; p. 14 krtphotos/Newscom.com; pp. 16, 21, 24 © Mary Evans Picture Library/The Image Works; p. 23 © C. Walker/Topham/The Image Works; p. 27 Hemera/Thinkstock; p. 33 © Mary Evans/Ronald Grant/Everett Collection; p. 38 David McNew/Newsmakers/Getty Images; p. 41 James P. Blair/National Geographic; p. 44 De Agostini Picture Library/Getty Images; p. 47 Hulton Archive/Getty Images; p. 49 Douglas Graham/Roll Call/Getty Images; p. 54 © Mary Evans/COLUMBIA PICTURES/EMI FILMS/Ronald Grant/Everett Collection.

Designer: Matthew Cauli; Editor: Andrea Sclarow;
Photo Researcher: Amy Feinberg